Roberto Rivas Valencia

Operation of a 4-Stroke Internal Combustion Diesel Engine

Roberto Rivas Valencia

Operation of a 4-Stroke Internal Combustion Diesel Engine

And its Influence on the Agricultural Tractor and Equipment, Thresher and/or Combine and/or Grain Harvester

ScienciaScripts

Imprint

Any brand names and product names mentioned in this book are subject to trademark, brand or patent protection and are trademarks or registered trademarks of their respective holders. The use of brand names, product names, common names, trade names, product descriptions etc. even without a particular marking in this work is in no way to be construed to mean that such names may be regarded as unrestricted in respect of trademark and brand protection legislation and could thus be used by anyone.

Cover image: www.ingimage.com

This book is a translation from the original published under ISBN 978-620-3-87571-3.

Publisher:
Sciencia Scripts
is a trademark of
Dodo Books Indian Ocean Ltd., member of the OmniScriptum S.R.L Publishing group
str. A.Russo 15, of. 61, Chisinau-2068, Republic of Moldova Europe
Printed at: see last page
ISBN: 978-620-4-16891-3

Operation of a Four Stroke Internal Combustion Diesel Engine and its Influence on the Agricultural Tractor, Equipment and Threshing and/or Combine and/or Grain Harvester.

Table of Contents

I. Introduction

Undoubtedly the four-stroke internal combustion diesel engine is the main supplier of power, so that agricultural tractors have movement and can develop the different activities in the field, depending on the implement to be used, and to make the fallow has to use a plow, either disc or moldboards, tracking with a disc harrow lift discs, or pull or pull, planting with a seeder, either normal or precision, likewise to make the crops, these should be done preferably with a cultivator, pest and disease control with a spray pump and harvest with a combined or harvester, if we use the above will advance efficiently, from land preparation to harvest, we will get crops in the shortest time, cheaper and higher quality.

Currently there are different capacities in terms of horsepower (Horse Power, H.P., diesel engines, and likewise, if they have more horsepower, more progress will have to use different agricultural implements including, (plow, harrow, planter, cultivator, trailer, dethatcher, mower or mower conditioner, windrower, baler, pruner, mill, grader, forage harvester, pump sprayer, combined or combine harvester), and that have a wider cutting width, will finish the agricultural activities in the shortest time, to this type of agricultural machinery and equipment while more power has a tractor and wider is the advance of the implement used, we will call efficient machinery and equipment, including the combined or grain harvester, because in a shorter time will perform agricultural work and its duration period, will be greater in terms of hours worked or years, and this is because they come more reinforced in terms of its structure, starting with the diameter and number of pistons or cylinders and all other parts that make up the engine, which we will be describing its operation.

Using agricultural machinery and equipment, a single farmer can meet the food and fiber needs of approximately 52 people.

189 years ago. Farmers used hand tools such as the sickle and scythe to harvest grain and in one day they harvested 8,160 m^2, which is equivalent to 00-81-60 hectares.

Objective: To know the operation of a four-stroke internal combustion diesel engine and its relationship with the operation of the tractor, agricultural equipment, grain harvester or combined or also called threshing machine.

II. Foundation

For a better study, we will start from the following regarding the systems that integrate the operation of a four-stroke internal combustion engine with four, six or eight pistons or cylinders and considering the basic parts that make up the operation of the engine.

1. - Four-stroke internal combustion system

2. - Cooling System
3. - Lubrication system
4. - Electrical system
5. - Injection system

Four-stroke internal combustion system: Regardless of the number of pistons or cylinders, for the engine to work properly, four strokes must be carried out, which we will call.

1.- Four-stroke internal combustion diesel system

    a. Admission
    b. Compression
    c. Explosion
    d. Escape

    a. Intake: It is the entry of air through the air filter that continues to the intake pipe and then to the piston chamber or cylinder which is descending, ie it is sucking air, the intake valve is open and the exhaust valve is closed.
    b. Compression: The piston or cylinder is rising, the intake and exhaust valve are closed, it is generating a high compression and heat.
    c. Explosion: When the piston or cylinder reaches the highest part, the compression that generates a temperature that exceeds 1000 ° C, the intake and exhaust valves are still closed, at that time the injector makes an application that generates a mixture of fourteen parts air to one part fuel and results in a considerable explosion, which generates the piston or cylinder necessarily down.
    d. Exhaust: Once the piston or cylinder begins to rise again, the exhaust valve opens and allows fuel residue and other impurities to circulate through the exhaust manifold and continue into the exhaust pipe and is the smoke you see when a diesel engine is running.

III. Function of the components:

Air filter: Its function is to allow the air to enter free of impurities or foreign substances, which affect the operation of the agricultural engine.

Intake pipe: Leads air into the internal combustion chamber, where the pistons are located.

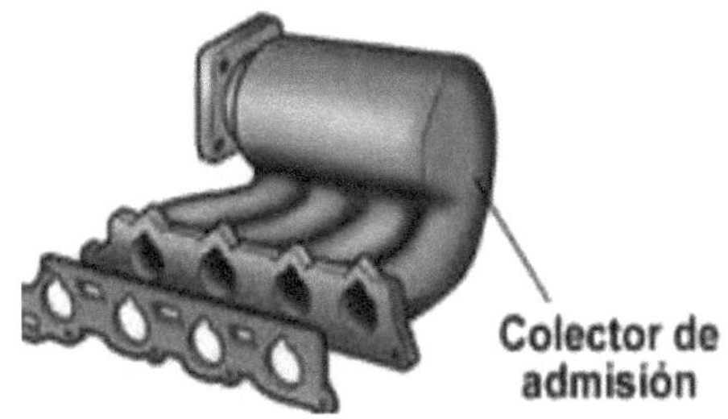

Piston or cylinder: It is a light steel structure, which is attached to the connecting rod and this is coupled to the crankshaft, the figure shows the connecting rod, the bolt or fastener, locks and slots, as well as metals or bearings and cover that is attached to the crankshaft with two screws.

The metals that allow the crankshaft to rotate at the base of the monoblock and the metals that facilitate the movement of the connecting rods, attached to the crankshaft, have a certain tolerance once the engine is repaired, when the engine is new, it is said that it is in standard, ie the metals have no wear, once the engine is repaired, due to wear, you have to install metals with measures in 10 or 15 mm thick and this is measured with platigage, is a strip of plasticine with its proper scale.

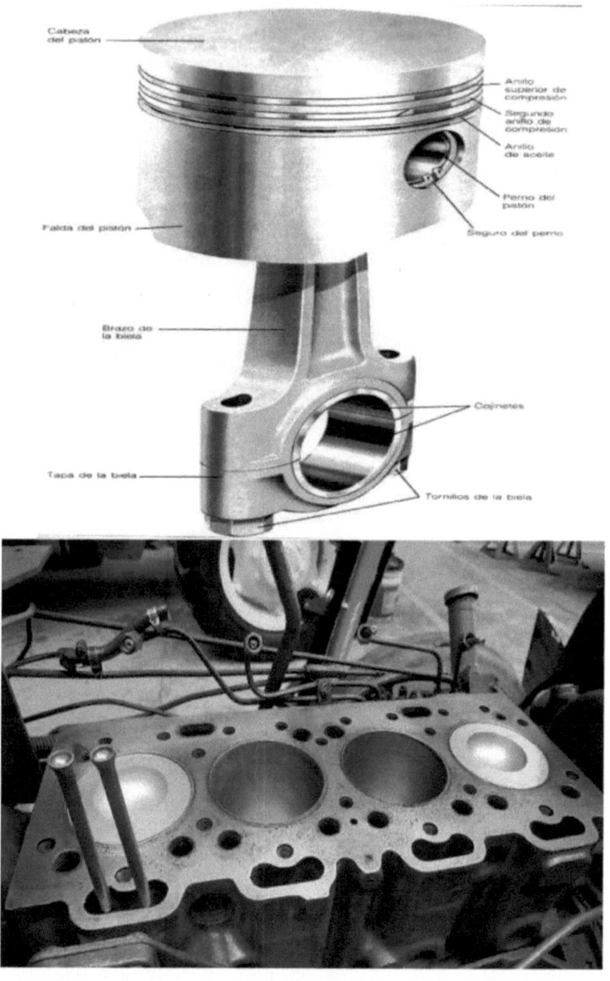

View of a four-stroke internal combustion engine, you can see the four pistons and liners, is where they make their journey and pipe where the diesel is driven.

Monoblock: Structure that results once the four-piston diesel engine is completely disassembled.

Crankshaft: Special steel structure and is the one that makes the pistons or cylinders move, is the main axis.

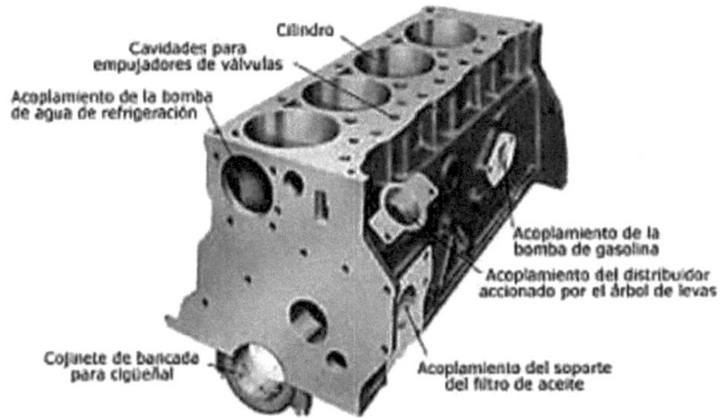

IV. Cooling System

Parts that make it up:
Radiator, hoses, water pump, fan, belt, water flow holes, thermostat, temperature indicator, taps.

Process: When the engine is on, because it is cold, the water begins to circulate inside the engine, through the hose from the bottom, which pulls the water from the radiator due to the operation of the water pump and is introduced by the ducts, which are designed to circulate the water, because the engine does not have the working temperature, makes the journey to the thermostat, which remains closed and does not let the water to continue, once the temperature

reaches 80 ° C, the tractor is working normally, but because in the work of the agricultural tractor is used a lot of power, the temperature begins to rise and when it reaches 90 ° C, to lower it has to automatically open the thermostat and the temperature begins to fall because the water is circulating to the radiator through the hose that comes out of the lid where the thermostat is located, once it reaches 70 ° C to 75 ° C is closed again and so the cycle continues.
So here it is suggested that the radiator is always supplied with water and antifreeze, the latter helps to stabilize the temperature.

Radiator: structure, called in other terms water reservoir

Hose to conduct the water: Its function is to conduct the water from the radiator to the interior of the engine and from the engine to the radiator, in the image in which the hose of return of the water to the radiator is observed.

Water pump: Its function is to pull the water from the radiator and impel it towards
inside the engine, so that the temperature stabilizes.

Fan and band: Its activity is to generate air through the bands and attract it to the engine, also so that the temperature is leveled and the bands or band whose structure is resistant rubber, which drives the fan, the alternator and the water pump, is driven by the rotation of the crankshaft through a pulley.

Thermostat: Device that regulates the outlet of the water from inside the engine, towards the radiator, once the temperature is at a maximum of about 90° C, and closes when the temperature drops to about 70°C to 75° C.

V. Lubrication system

Components that make it up:

It refers to all parts of the engine that must be lubricated, crankshaft, metals, connecting rods, pistons, oil pump, liners, gears, monoblock walls, camshaft, rod bases, rods, rocker arms, springs, spring locks, spring retainer caps, exhaust and intake valves, the ducts through which the oil circulates and the half-moons that are located between the crankshaft and the bedplate.

The engine starts working, the oil pump begins to work, which sends it to the parts that are in motion, such as the crankshaft, the metals of both the crankshaft and the connecting rods, the pistons, the piston rings, the oil pump itself, the gears of both the oil pump and the camshaft, the rod, the rocker arms, the springs, locks and valve covers, both intake and exhaust, for the gears of the oil pump and the camshaft, the gears of both the oil pump and the camshaft, the rods, the rocker arms, the springs, locks and valve covers, both intake and exhaust valves, for the gears that are in the front timing cover, the gears or mechanisms of the injection pumps.

So the oil has five basic functions (Hunt, 1983): Stabilizes the temperature, by the fact that it has a certain viscosity, reduces wear and friction by separating the parts in friction, exerts a washing system, once the engine was turned off, due to the viscosity and that by gravity will have to go down during the time that the engine is off, this will drag to the lowest part which is the crankcase, all impurities that is found in its path in the descent, absorbs the shocks that occur in the bearings and acts as a seal in the combustion chamber.

Oil pan or reservoir

Oil pump: Device that absorbs the oil and impels it to the parts that are in contact or friction inside the engine, such as crankshaft, bedplate and metals, pistons, connecting rods and their metals, rods, rod bases, camshaft, gears, liners, rocker arms, head springs, valves and other components that are in constant movement and friction.

Sleeves: Cylindrical structures, of a resistant plastic, which are used for the piston to make its journey, in their spaces, have rubber bands or gaskets, which do not allow water to enter the combustion chamber, and is where the piston makes its journey, up and down and lubricated with oil, so that if the bands are not working, the oil is mixed with water and the engine can desbielarse, ie stops working.

Gears: Round toothed structures, which make other parts work, linked to other gears. ________________ ____________________________

Camshaft with gear: Structure that works around bushings and a gear makes it rotate through the bases or bushings, these drive the rods that go up and down, and at the same time the rods drive the rocker arms and these determine that the intake and exhaust valves open or close. Bushings: Round structures, which serve as a base for the camshaft to rotate.

Rod bases or, pusher: It is observed as the camshaft rotates clockwise and the highest part of it pushes the rod and makes the rod go up and down or up to open or close the intake and exhaust valves with the help of the rocker arm, in this case as the spring which is observed has enough pressure automatically when the rod is going down, the spring makes the intake or exhaust valve to close, the spring is fixed on the valve by a cover, which fixes the valves by means of wedge-shaped locks, which already have their space between the tip of the valves and the cover or lid.

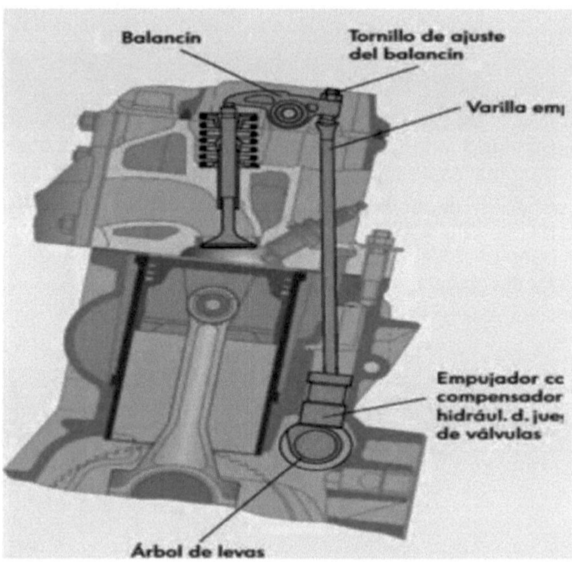

Rods: Elongated structures, at the bottom round and at the top with a recess and that the camshaft makes them work, pressing by means of the rocker arm to push down, to open and close the exhaust and intake valves.

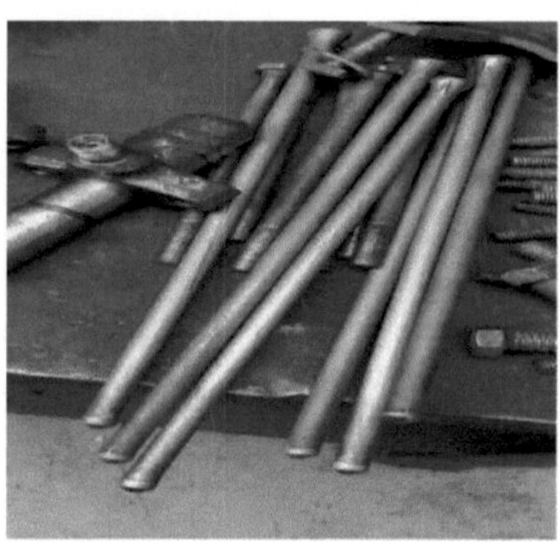

VI. **Electrical system**

Components: battery, cabling, starter or starter motor, engine flywheel, generator or alternator, regulator, ignition system, choke, fuses, electrical charge indicator or ammeter, headlights.

When activating the engine ignition, the electrical energy stored in the accumulator makes the bendix work, which is located in the starter motor or gear and this is inserted into the slots of the steering wheel which will rotate the crankshaft and all internal parts, turning on the internal combustion engine to four times, a red power cable, is the one that supplies the electrical energy and another black color that is the land is attached to the chassis of the tractor, the cables are distributed in the network of the tractor and some are connected to the headlights, dashboard, preventive and a cable is connected to the generator, which is responsible for generating electricity, this cables come out and one of them is connected to the regulator, which provides some energy to the accumulator or battery, to have enough energy, if I want to turn off and turn on the engine or turn on the lights, while the engine is off.

Battery: Structure where electrical energy is stored, and is available for when the engine is started and the lights, or turn the key to activate the current, you can turn on the lights or preventive, also to know how much fuel (diesel) contains the diesel tank.

Battery, cables, gear and connection to the engine ignition: The battery is observed, it leaves a black wire, which is the ground and is fixed with a screw to the tractor chassis, another red wire which is properly the current and is connected to the gear or starter motor, this leaves another cable which is linked to the alternator or generator, from there cables to the lighting system, the ignition and the regulator, which determines the amount of electrical energy is going to enter the battery.

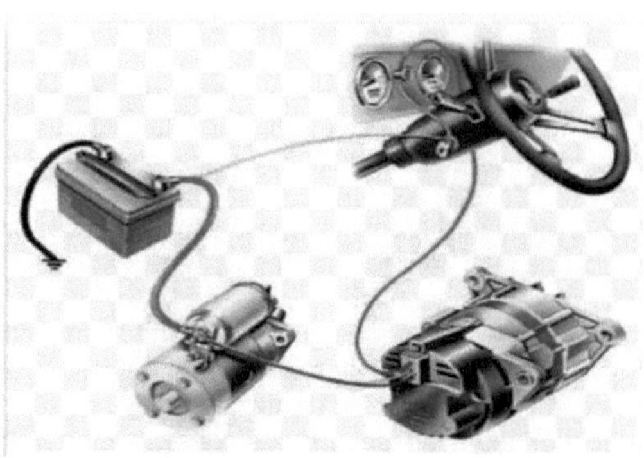

Engine flywheel: As part of the ignition system and which is inserted and bolted to the back of the crankshaft.

VII. **Injection system**

Components: Fuel tank, hoses, filters, lift pump on some engines, example Perkins brand, injection pump, tubing, injectors, recoil tubing, on some engines, example Perkins brand.

When the engine starts, the fuel begins to flow through hoses or tubing from the fuel tank (diesel), to the filters, then to the lift pump, to follow its path in the direction of the injection pump, which processes the fuel (diesel), and projects it to the injectors, which apply it in each of the combustion chambers, where the pistons or cylinders are located, to take the time of explosion. The excess fuel and that is not possible to apply the injectors, is returned to the injection pump and the cycle is repeated.

Functions of the components of the injection systems:

Fuel tank: Space where diesel fuel is stored.

Flexible pipes: They allow the fuel (diesel) to circulate from the fuel tank to the injectors, which make the fuel pulverization in the third period called explosion, of a four-stroke internal combustion engine.

Diesel and water filter: They allow us to retain impurities and water, and that the diesel is circulated and used as clean as necessary, so that the fuel system works in the most efficient way.

Lift pump: It is responsible for driving the diesel fuel to the injection pump.

Four-piston, four-injector injection pump: It is responsible for processing the diesel fuel and propelling it to the injectors, by means of a pipe.

Injectors, rods, toe cap, screws, nuts and washers: They are the structures responsible for applying a part of the fuel and that is mixed with 15 parts of air to one part of fuel, so that the explosion takes place and this makes the piston necessarily down with force.

Head armed with exhaust and intake valves, springs, caps and locks, even though, the image that is presented is 4 exhaust valves and 4 intake valves, the general rule is that they are 8 intake valves and 8 exhaust valves.

VIII. **Horsepower of a four-stroke internal combustion diesel agricultural engine.**

In this section, we will talk about the horsepower or H.P. (Hourse Power), and generated by the pistons, which are those that at a time provide the power of the diesel engine, and so we can say that the power is then the work done by the unit of time.
The steam engine was first invented in England and in the latter part of the 18th century, James Watt determined the capacity of his steam engines in terms of horsepower.

He conducted a series of tests on average horses and found that one horse could extract 366 lb of coal from a mine with a speed of 1 ft/sec. In other units this was in round numbers 22,000 ft*lb/min. And arbitrarily James Watt considering that the material of the steam engines was iron increased by 50 % to that already calculated, obtaining 33 000 ft*lb/min., operation that develops in the following way 22,000 * 1.5 = 33, 000 ft*lb/min. And that this is equal to 550 ft*lb/sec, i.e. already converted to seconds.

There are two systems for measuring diesel engine horsepower.

1. - International System (SI), in which the unit of power is called watt (W). One watt is the power equivalent to one newton of force applied across one metre of distance in one second.

A newton (N) named after Sir Isaac Newton is the unit of force required to accelerate one kilogram of mass by one meter per second every second.

2. - System é(SI), in which horsepower (H.P.) is considered as a result, one pound of force in the English system is approximately 4.448 N. One horsepower in the English system is equivalent to 745.7 W, and one kW (KW) is equivalent to 1.341 H.P. and 1 W is equivalent to 1000 kW.

Examples:

Assume we apply a force of 100 N [22.48 lb] power is applied and that is equal to force times distance times time and with a velocity of 4 m/sec [13.123 ft/sec]. [13.123 ft/sec].

Let's find the necessary power, in both types of systems.

Power in the International System:

400 W = 100 NX4 m/1 sec*1 W/1 N*m/sec

So if one Watt is equal to 1000 Kw, we have that 400 W/1000 kW = 0.400 kW so if we multiply 0.400 kW by 1.1341 H.P., according to the conversion already explained, we would get the following 0.400 kW * 1.341 H.P. = 0. 536 H.P. which is the equivalent to the English System.

Power in English system: 0.536 H.P. = 22.48 lb * 13.123 ft/1 sec*1 H.P./550 ft*lb/sec.

If we need to know in a practical way how much fuel a tractor spends per hour, we fill the tank completely, then we work an hour in the necessary activity in the field, after the hour worked, we fill the fuel tank again and what we fill, will be what I spend per hour. We can also measure 100 linear meters and we take time, if the tractor travels for 2 minutes, by means of a rule of three we obtain 1000 linear meters that are one kilometer and we have 20 minutes and by means of a rule of three we convert them into hours and we get 3 kilometers per hour.

The power in an agricultural tractor is based primarily on the diameter or volume of the cylinder or piston, the larger the diameter or volume, the greater the power and also the more cm3 more power, or put another way in cubic centimeters (cc) or liters, or displacement.

If we are talking about a new tractor, we are saying that the engine is also new, therefore, we can say that the engine is in its standard state, i.e. it has no wear at all.

Therefore, it is suggested, to use an engine of 95 H.P. to the steering wheel, which will carry out the activities of efficient form, also it is considered to use implements with a greater width of advance or of cut and will be obtained greater capacity of hectares per hours, carried out in the day of daily work.

IX. Repair of a four-stroke internal combustion diesel engine, e.g. 4, 6 or 8 pistons.

The repair of a deviated engine or that by heating by an error is completely damaged the diesel engine of four, six or eight pistons or cylinders, which are the most common in agricultural tractors and combined or harvesters. We proceed to disconnect the engine from all the parts that connect the engine to the tractor, the box that protects the engine, remove the water through the tap or radiator key to disconnect the water inlet hose that leads to the inside of the engine and the hose that allows the return of water from inside the engine to the radiator and remove the four screws that secure the radiator to the base designed for this function to separate the radiator, then disconnect all hoses, cables, pipes that are connected to the engine and supports that serve to fix the engine to the chassis of the tractor, also remove the crankcase screw to remove the oil from the engine.

Once the engine is completely disconnected from the tractor, we proceed with a structure either with a tripod or with some other structure to lift the engine of the tractor, and proceed to move and deposit it on the floor.

Once on the ground the next thing to do is to disassemble the engine with the different tools, astro or Spanish wrenches, ratchets, sockets, dies, hoses and tubes etc.

Once separated the engine components, such as crankcase or oil tank, oil pump, crankshaft, pistons, rings, connecting rods, camshaft, bases or bushings, rods, shirts, rocker arms, head, valves, valves, springs, valve locks, rocker cover cap, water pump, cover and thermostat, different gears, pulleys, fan and base, engine flywheel and clucth, fuel pump, injectors. We would be left with the main structure which would be the monoblock and of course all the bolts and nuts that joins all these parts to the engine.

Once the engine is disassembled, proceed to wash all parts with oil preferably, because it is less expensive and flammable in relation to gasoline, which will be used in certain cases, such as washing the crankshaft, connecting rods, oil pump, this is if warranted.

In the case of the monoblock and heads once disassembled and washed with oil, are impregnated with a substance containing acid and that is the carbonf, the time that is left submerged both the monoblock and the head from one day to another, and this to completely detach the oil residues stuck to the walls, residues that with the washed with oil is not achieved.

Once they are clean, they are sent to a specialized rectification workshop, if the damage consists of a devielada for lack of oil or lack of power, due to use, the workshop will carry out its activity and through a review, proceed to polish both the stumps of the crankshaft, and the bench where the crankshaft is located and the same will be done in the case of the connecting rods, polish the crankshaft journal and the part of the connecting rod where the metals are located and determine the wear due to the work in hours worked and suggest if the metals that have to be purchased and placed in the corresponding part that in 10, 15 or 20 mm, then they will have to be replaced, 15 or 20 mm, then the metals will have to be placed according to the recommendation of the workshop specialized in rectification, because it is the space that results between the metal or bearing and the crankshaft journal, the same is the space to be considered between the space of the metal or bearing of the connecting rod and the crankshaft journal, also will change the round metals that allow turning the camshafts and at the same time the camshaft will be checked to see if it is not damaged and if so it will have to be polished or changed.

And in the case of the head, the seats and intake and exhaust valves will be changed, which are a total of 16 valves, for each piston or cylinder are two exhaust and two intake.

Immediately, since it was received by the workshop specialized in rectification of the monoblock, crankshaft, connecting rods and pistons, we proceed to clean them and the crankshaft must be adjusted as follows, the metals are placed on the bedplate, space where the crankshaft is placed and the metals of the covers or bearings and placing both the metal of the base of the crankshaft, as the metal of the bearing or cover to fix the crankshaft a strip of plasticine that comes in an envelope called plastigage, which brings a graduation, so that once applied a pressure of 120 pounds to the bearings or covers or covers of the crankshaft, the covers are removed again and the correct space is 2 to 3 mm, has this space the crankshaft is adjusted, In the case of the connecting rods the space is 1 to 2 mm of space between the metal of the connecting rod which carries a pressure of 60 pounds and the crankshaft journal, so that they are already adjusted, both the metals of the connecting rods, as the metals of the crankshaft.

Next, you also have to check the oil pump and change the spare parts, which can be seals, springs and gaskets. This is in order that the pump generates the proper pressure for the engine to function in the best way, as it is verified that the pump has adequate pressure. Once the spare parts were changed, the part where the oil enters the pump is placed and this part is placed on the tongue and the gear is rotated manually and feel the pressure, if the pressure is little proceeds to change to another shorter spring or modify the one already purchased, so that the oil pressure entering the oil pump and at the same time expels the parts where it is required.

In the case of the head, as it comes with new seats and new valves also new, you have to seat the valves both the intake and exhaust, well once the head is clean, we proceed to place solid emery where it is supposed to be the part that will make contact with the seat of the head and with a sucker for valves, this is inserted into the flat part of the valve and proceeds to turn up and down the seat of the seat of the head and when they have spent between 10 and 15 minutes of continuous work of the 16 valves, the head is assembled, first placing the cap or rubber on the inside of the head and placed on the base of the valves, in addition to its spring corresponding to each valve, its wedges or insurance and its cover, we proceed to use the arc, which contracts the valve and place the insurance or wedges in the little channels that have the valve around and the cap or cap, this is in order to separate the wedges or insurance and caps or caps and do not move the valves.

Well, then the head will be tested for leaks, therefore, you have to introduce gasoline, because it is thin in each of the holes leading to the intake and exhaust valves and verify that the gasoline does not leak where the intake or exhaust valve makes contact with the seat of the head.

Of course, gears or removable metal parts should be checked for physical condition and replaced if necessary.

Similarly you will have to check the thermostat, to see if it is opening at the right temperature, you can try if this is immersed and held with a wire in a jar with water, which is applied heat by coal or wood and a temperature indicator can be checked if you have to open at a certain temperature, for example in the case of agricultural engines, should open at 80 ° C if it is observed that does not open, this indicates that it does not work, it is best to change it for a new one.

Another part that should be checked and if necessary replaced is the water pump, which contains seals and the propeller and if necessary replaced with a new one.

Well, once you have checked the above, when you are assembling the engine, you have to change the corresponding gaskets, which can be of asbestos or cork and these can be impregnated with oil or shellac glue if applicable.

Once the pointers are installed, they have to be calibrated and the order for this function is 1-3-4-2, that is the piston number 1 and 3 must be in the dead center, it is the highest part of the piston 1 and 3 and proceed to calibrate both the intake valves (0.2 mm) and the exhaust valves (0.5 mm), and in the next phase the valves that integrate the pistons 4 and 2, in the same way.

X. **Four-stroke internal combustion diesel engine operation**

Let's assume that all engine systems are in normal condition, then the following happens:

If the engine is switched off by the choke, the choke must be in the open position, so that the fuel (diesel) can circulate.

Well, when you turn the key in the switch or ignition device, immediately electrical energy comes out of the battery and drives the bendix of the march, through the red cables that of the positive charge, but also has an important role the black cable which is the ground and is connected to the chassis of the tractor, well then the bendix is activated and connected to the engine flywheel through some astrias and starts the engine activity in this sense it is said that the current is flowing to the parts where it is needed, the dashboard, lights, and this energy is generated in the alternator or generator, and this energy is being generated in the alternator or generator from there a cable goes to the regulator which allows some electrical energy is stored in the accumulator, also begins to flow the oil, the crankshaft is turning clockwise, this makes the pistons develop the four strokes already explained and for this it is also necessary that the camshaft is working, to make the rods go up and down, which are in the bases or buses, up and down and this allows the rocker arms are working, and previously calibrated and allow both the intake and exhaust valves open and close when appropriate, on the other hand the water that is in the radiator is being introduced into the engine and makes its way in order to stabilize the temperature, due to the power that is applied, the temperature tends to rise, and when it is located at 90 ° C, as it is a temperature that if it continues to increase can affect the engine, then at that moment the thermostat opens, which allows the water that is inside the engine to circulate freely to the radiator, in order that the temperature drops to approximately 70 to 75 ° C and immediately closes the thermostat and repeats the cycle, At the same time the diesel fuel leaves the tank and through pipes and by the action of the lift pump, diesel moves, but before it reaches the lift pump passes through at least two filters, one as a water trap and the other as a trap for impurities, well the fuel leaves the lift pump and is directed to the injection pump through copper pipe, there is processed in such a way that also through pipes expels the diesel to the injectors, which apply a certain dose of diesel in the time of explosion.

So we can say that the engine transforms electrical energy into mechanical energy, well we will also mention that the engine flywheel is turning, and suppliers of agricultural tractors as an example the brand of agricultural tractors Kubota Uruapan (12 March 2021), consider that the model M9540DTH-MEX generates a net power to the flywheel of 95 H.P. and in 70.8 in kW, it means that even the flywheel of the diesel engine has up to 95 H.P., and we will explain how the force or power is transmitted to the rear of the engine, and up to the application of the power in the power take-off and drawbar and of course, depending on the power, so will be the capacity of the hydraulic system to lift and lower implements.

XI. **Maintenance of the four-stroke internal combustion diesel engine and a tractor in general. For the above, it is based on the Ford 5600, 6600 and 7600 tractor.**

Every 10 hours of work, daily

Check the oil level, if the engine is hot, let it cool down for 15 minutes, remove the dipstick from the oil gauge and if it is low, top up to the maximum.

Pre-filter rate, loosen knob, remove cup and clean

Check the water level in the radiator, if it is low, top up with antifreeze solution.

Oil bath air cleaner, remove radiator grill. Loosen knob and remove cup, drain and clean cup and old cup, if dirt reaches a level of 9.25 inches (6 mm). Fill the cup cup to the level mark with clean engine oil.

Air cleaner dust collector, A rubber dust collector is located under the air cleaner body. Periodically squeeze the collector, which will open and discharge the accumulated dust, if the air cleaner restriction warning light comes on while the engine is running, loosen the screw and remove the cover, loosen the knob and remove the outer filter element. Clean the filter by tapping it with the palm of your hand. It should not be tapped against a hard surface as this may damage it. It can also be cleaned with compressed air at less than 30 lbs/in2 (2 Kgs/cm2). Insert the nozzle through the inside of the filter and blow the dust out. Particles stuck to the outside should be blown out with the nozzle separated from the filter by at least 6 in. (15 cm).

Every 50 hours

Check and inflate the tyres to the correct pressure, check for possible damage to the tread or sidewalls.

Adjusting the clutch pedal

Check the oil and if necessary top up the threshing pulley, if fitted.

Check and top up the battery electrolyte level, if the battery is sealed this check is not done.

Open the fuel filter drain plugs to expel water and impurities.

Check front wheel pressure and rear, according to the following table:

	82 lbf/foot (11 mkg)
Front disc to hub nuts (no four-wheel drive)	
	240 lbf/foot (33 mkg)
Front disc to hub nuts (with four-wheel drive)	
	130 lbf/foot (18 mkg)
Front disc to rim nuts (with four-wheel drive)	
	320 lbf/foot (44 mkg)
Rear Disc to Hub Nuts (with manual adjustment wheels)	
	130 lbf/foot (18 kmg)
Rear disc to rim nuts (with manual adjustment wheels)	
	252 lbf/foot (73 mkg)
Rear Disc to Hub Nuts (with mechanically adjustable wheels)	
	275 lbf/foot (38 mkg)
Rear Rim Clamp Nuts (with Mechanically Adjustable Wheels)	

Combined transmission/rear axle level (without handbrake to transmission)

With all the outer cylinders extended, check that the oil reaches the "FULL" mark on the dipstick, if necessary top up through the filler cap.

Combined transmission/rear axle level (with handbrake on the transmission), if the optional handbrake is fitted to the transmission, there is a filler plug instead of the dipstick. Fill via the filler plug.

Front axle hub oil level (Four-wheel drive only), place the left front wheel on the horizontal line, remove the combined fill level plug and fill the hub, replace the plug, do the same with the right rim.

Differential oil level (4-wheel drive only), make sure that the oil level reaches the filler opening/level. If necessary top up and refit the plug.

Front wheel hubs (without four-wheel drive), in difficult working conditions grease daily.

Greasing the brake pedal pivot

Grease the clutch pedal pin

Front wheel hoses (without four-wheel drive), one grease nipple each side
Grease the front axle center pin (with four-wheel drive or power steering).

Grease the front axle universal joint (four-wheel drive only), left and right side.

Grease the front axle universal joint (four-wheel drive only), left and right side.

Grease the universal joint on the front drive shaft (four-wheel drive only).

Grease the universal joint on the rear drive shaft (four-wheel drive only).

Grease the leveling box and hydraulic lift linkage.

Grease the latch of the automatic coupling (if fitted).

Every 100 hours

Changing the engine oil, it is suggested to warm up the engine to its working temperature. Stop the engine and remove the drain plug and collect the oil in a suitable container.

Remove and discard the oil filter, clean the block surface and reassemble a new filter, not overtightening the filter.

Run the engine for a few minutes to distribute the oil before checking the level with the dipstick.

Every 150 hours

Change the oil only, there is no need to change the engine oil filter.

Every 300 hours

Change engine oil and filter

Roll bar or safety cage (if fitted), check all bolts and nuts for tightness.

Hydraulic power steering, with the front wheels in a straight line, remove the cap and fill the reservoir until the oil level reaches 0.75

in. (19 mm) below the bottom of the filler neck.

Hydraulic oil filter, remove and discard the filter. Clean the manifold face before fitting a new filter. Do not over tighten the filter.

Fan belt, the belt tension is correct when the deflection is 0.50 -0.75 inch or (13-19 mm), to adjust, loosen the bolts and move the alternator.

Drive belts, to adjust the fan belt, loosen the bolts and move the idler pulley until the belt flexes 0.50 to 0.75 in (13-19 mm) in the center of its longest travel. (13-19 mm) at the center of its longest travel to adjust the alternator belt, loosen the bolts and move the alternator until the belt can flex 0.50 to 0.75 in (13 to 19 mm).

Injection pump, Drain the oil, removing the filler plug, level plug and drain plug, place the drain plug and fill the pump with clean engine oil until it comes out of the level plug, place the filler level plugs.

External air filter element (if fitted), Take the air filter out, wash the filter and clean it with hot water and a little detergent, rinse the filter in clean water, shake off the excess and let it dry by itself.
Once dry check its condition by inserting a light, it should pass through evenly. If a light beam is visible, the filter is perforated and cannot be used, clean the filter housing before reassembling and covering it.

Adjusting the foot brake, raise the right wheel and place a 1 to 1.5 inch (38 mm) block between the right pedal and the bottom of the platform, loosen the locknut and turn the adjuster to lock the wheel. Tighten the locknut, remove the block and repeat the operation with the left wheel and pedal. Then check that the brakes are balanced.

Adjusting the parking brake (if fitted), loosen the lock nuts and remove the pin from both actuating rods. With the handbrake rod released, push down the two cross shaft arms until they touch the "X" brake pedals, adjust the length of the rods by turning the forks until the pins can be inserted into the bottom of the slot in their respective forks, secure the pins with cotter pins and retighten the rod locknuts, test the handbrake to make sure it is working properly.

Adjusting the parking brake to the transmission (if mounted), tighten the self-locking nut until 30 to 45 pounds (13.6 to 20.4 kilograms) of force must be applied to the handle to engage the first sector slot.

Service every 600 hours, carry out the above checks plus the following:

Air cleaner in oil bath (if fitted), remove precleaner, extension tube and radiator grille, loosen screws and clamps and remove air cleaner, clean inner cup, cup and mesh in solvent, allow to dry and insert air cleaner body, fill both cups to correct level with clean engine oil and place in air cleaner body, refit radiator grille and precleaner.

Air cleaner outer element (if fitted), replace the outer element every 600 working hours or after 10 washes, whichever comes first, loosen the screw and remove the cover, remove and discard the outer element after removing the fixing knob, clean the inside of the filter housing with a damp lint-free cloth, fit the new filter and fit the cover.

Steering box oil level (Manual steering only), remove the left cover of the steering gear to reach the steering box, fill with oil up to the oil level plug.

Front wheel bearing (No four wheel drive), pull hand brake, lift one front wheel. Remove the wing pin plug, nuts, thrust washer and outer bearing.
Remove the complete wheel and remove the grease retainer and inner bearing, clean the parts in solvents and allow them to dry, inspect the bearings and bearing cups for signs of discoloration or wear, fill the space between the two bearing cups with grease, grease the sleeves, mount with a new grease retainer and tighten the castellated nut to 20 to 30 pounds/foot (2.77 to 4.15 kilogram meters), turn the wheel hub clockwise three to six times, retighten the castellated nut to 45 to 55 pounds/foot (6.22 to 7.60 kilogram meters), loosen the nut two notches and then tighten just enough to line up with the hole in the stub axle, install a new pin and replace the plug, repeat the procedure on the other wheel.

Fuel injectors, loosen the pump to injector tubes at their connections, clean around the injectors and remove the pump to injector tubes and the overflow tube, discarding the copper washers on each side of the connections under the overflow inlet, if you do not have a spare set of injectors, cover the ends of the pipes and the inlet and outlet of the injectors to prevent dust from entering, remove the injectors after loosening the nuts, remove the copper washers from the housing of each injector and the cork dust cap and discard them, Fit the replacement injectors with new washers and dust caps, reconnect the surplus pipe with new washers to each side of the low connection and tighten, reconnect the pump pipes to the injectors tightening only the pump terminal, once the injectors have been fitted bleed the system, refer to draining the filters and flushing the injection system, the injectors that have been removed will be handed to an authorised Ford tractor agent for reconditioning and subsequent use in the next 600 hour service.
Fuel tank filter, close fuel tap (if fitted) by turning clockwise, loosen screw to remove filter element and cup, flush cup with clean fuel and fit new element and gaskets, open fuel tank tap and bleed system.

Bleed the fuel system (distributor type injector pump only), loosen the bleed screw on the top of the filter and let the diesel out until it comes out bubble free, tighten the bleed screw and loosen the bleed screw on the injector pump, with the control pulled out, turn the engine until the diesel comes out bubble free through the bleed screw hole, tighten the bleed screw of the injector pump, loosen the connections of the injector pipes and keep turning the engine with the stop control pulled and with the throttle at full throttle until the fuel comes out bubble free, tighten the connections one by one while the engine keeps turning,

Bleeding the fuel system (In-line injector pumps only), loosen the filter bleed screw and operate the priming lever on the injector pump, until the diesel comes out of the bleed screw hole bubble free, tighten the bleed screw, loosen the injector pipe connections and crank the engine with the stop control pulled and the throttle wide open until the diesel comes out of the connections bubble free and tighten the connections one by one while the engine continues to crank.

Rocker arm clearance, with the engine cold, place cylinder number one (front) at top dead centre on the ignition stroke through the inspection window to locate the P.M.S. (top dead centre), on the ignition stroke the two valves of cylinder number 1 will be closed, with a feeler gauge check the clearance between the valve stem and the rocker arm, turn the screw to adjust the clearance, check and adjust the following: Valves

Number 1 Admission	Number 2 Admission
Number 1 Escape	Number 3 Escape

Turn the engine one complete revolution until cylinder number 1 is at the P.M.S. on the exhaust stroke, (The valves of cylinder number 1 will be slightly open, with the intake valve opening and the exhaust valves closing).

Check and adjust the following, remaining, valves:

Number 3 Admission	Number 4 Admission

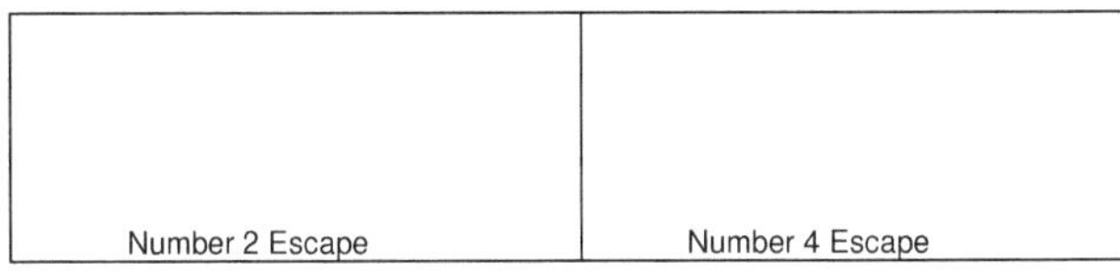

<table>
<tr><td></td><td></td></tr>
<tr><td align="center">Number 2 Escape</td><td align="center">Number 4 Escape</td></tr>
</table>

Replace the rocker cover with a new gasket or gasket (a), if necessary.

Service every 1200 hours

Carry out the above checks, plus the following:
Fuel filters (In-line dual filters only), the secondary filter must be changed, this operation must be carried out by an authorised tractor dealer.

Transmission and rear axle oil (All models), remove drain plugs, with two speed PTO (Power Take Off) option, also remove additional drain plug, replace plugs, once all oil has drained out.
Front axle transfer case (If fitted), remove drain plug, refit once all oil is drained, refill transmission/rear axle through filler plug until oil reaches correct level.
In the case of tractors without transmission handbrake), if the optional transmission handbrake is mounted, instead of the dipstick there is a level plug, fill the system until the oil reaches the level plug.
Differential oil (Four wheel drive only), remove the level/fill plug and drain plug and allow the oil to drain out, replace the drain plug and refill the diff case through the level/fill plug opening until the oil reaches the bottom of the opening.
Front axle hub oil (Four wheel drive only), position the left wheel so that the drain plug is at the bottom, remove the plug and allow the oil to drain out, refit the drain plug and position the wheel so that the line is horizontal, remove the level/fill plug and fill the hub to the bottom of the opening, refit the level/fill plug. Repeat on the right wheel.
Threshing pulley, drain the oil through the opening of the filler/level plug, removing the pulley and reversing its position, refill through the same filler/level opening.
Power steering oil filter (if fitted), Disconnect the hoses, which will drain most of the oil, remove the screw and tilt the reservoir back, discard the filter and O-ring and clean the pump and reservoir, fit a new O-ring and filter and fit the reservoir, fit the oil hoses and fill the reservoir with fresh oil until the level reaches 0.75 inches (19 mm), below the bottom of the filler neck, with the engine running, bleed the system, turning the handwheel from one end to the other several times, re-level.
Alternator brushes, replace the alternator brushes and clean the brush box, it is recommended that this service is carried out by a tractor dealer. Lift up the retaining clip and remove the plug and flat plug from the rear of the alternator. Remove the alternator.

Using a 2BA socket wrench, remove the fixing screws from the rear cover and remove it, leaving the fuse box visible. Noting the position of the wires, remove the screws and remove and discard the brushes. Lift the brush box, after removing the fixing screws.
Clean the collector ring and the inside of the brush housing with a clean, lint-free cloth.
With new brushes, fit the alternator in reverse order to dismantle it.
Service every 2400 hours
Carry out the above operations, plus the following operations
Radiator fluid, open the tap and drain the radiator, remove the radiator cap to increase the rate of fluid flow, open the tap and drain the engine block, flush the system and refill with a solution of 50% antifreeze and 50% water, this allows a two-year protection against frost and corrosion.

General Maintenance

Hand throttle adjustment, with the throttle lever at minimum, disconnect the throttle linkage from the crank, set the correct minimum engine speed by means of the minimum stop screw on the distributor type pump or by the stop dial on the in-line injection pump, loosen the lock nut and adjust the length of the throttle linkage to exactly match between the injection pump and the connecting rod, with the throttle lever at minimum, if the optional foot throttle is fitted, set the hand lever to maximum and adjust the compression spring adjuster so that the maximum speed at the pump can be avoided. Adjust the compression spring length to 4.75 inches (12.06 mm) by means of the adjuster, reset the throttle lever to minimum and disconnect the throttle linkage from the linkage, loosen the lock nut and adjust the throttle linkage to enter the linkage with lever fully forward, make sure that maximum and minimum speeds are obtained with the throttle lever and pedal.
Minimum motor speed (Final injection pump), Loosen the lock nut and turn the stop screw to set the minimum speed, the maximum no-load speed must be set by a specialist agent.
Adjustment of the injector pump (distributor type injector pump), the adjustment of the injector pump is made in the factory by means of special tools, if it is necessary to remove or change the pump, note the position of the line on the pump body, in relation to the adjustment marks on the front plate of the engine, mount the pump in the same position, to maintain the original factory setting.

Adjusting the injection pump (In-line injection pump), remove the flywheel access cover and turn the engine until the adjustment marks appear, using a screwdriver, continue to turn the flywheel clockwise until the arrow, If the shaft at the end of the camshaft is between the 26° and 28° setting marks (Ford 6600) or between the 24° and 26° setting marks (Ford 7600), remove the camshaft cover from the pump, if the shaft at the end of the camshaft is clear of the setting mark, then piston number 1 is on the exhaust stroke. Rotate flywheel motor one full turn until flywheel shaft is between 26° and 28° marks (Ford 6600) or between 24° and 26° (Ford 7600) with number 1 piston on compression stroke, check the setting by placing a pointer inside the V-notch on the end of the camshaft. The pointer should line up with the setting mark if this is not the case, proceed as follows, drain the radiator fluid and remove the lower hose, remove the injector pump setting cover from the front of the engine so that the pump gear is visible, loosen the three screws on the injector pump control gear and turn the gear adapter until the setting marks align, retighten the gear screws and fit the cover and gaskets, fit the lower hose and refill the radiator.

Replace the headlamp seal unit (if fitted), remove the radiator grille, separate the headlamp seal unit from the rubber headlamp, fit a new unit following the reverse of the dismantling procedure.

Headlight bulb replacement, integral (if fitted), remove the radiator grill, separate the bulb retaining spring and remove the bulb, fit a new bulb and install the bulb following a reverse process to dismantling.

Changing the side headlight bulb (if fitted), remove the clamp and the lens and reflector assembly, take out the lamp holder and change the bulb, fitting a reverse process to that of disassembly.

Changing the bulb in the instrument panel, the panel illumination warning bulbs can be removed from the back, to do this, remove the four screws and lift the instrument panel, if necessary, disconnect the hour meter cable.

Fuse replacement, the type and quantity of fuses available depends on the electrical equipment installed, the protected circuits are as follows:

Lighting circuits: One 15 ampere cartridge air fuse located behind the tank enclosure.

Windscreen wiper fuses; two 15 ampere bayonet overhead fuses to the right of the tank enclosure.

Light and turn signal equipment:
Two 8 amp fuses and one 5 amp overhead cartridge fuse located behind the tank casing, the 8 amp fuses protect the right and left lighting circuits respectively, the 5 amp fuse protects the turn signals.

Fuse box, lighting, turn signals, wiper motor and stop light (if fitted), located on the right hand side of the tank casing, there are six numbered fuses, which protect the following circuits.

Fuse number	Intensity	Circuit
1	8 amps	Road lights
2	8 amps	Dipped headlights
3	8 amps	Left front and rear position lights.
4	8 amps	Right front and rear position lights.
5	8 amps	Turn signal and windshield wipers
6	8 amps	Stop lights and horn

If an emergency warning device is included, an eight fuse box is installed. Circuits 1 to 5 are the same as the six-fuse system.

The following fuses protect the following circuits:

Fuse number	Intensity	Circuit
6	8 amps	Brake lights
7	8 amps	Horn
8	8 amps	Emergency warning lights.

XII. General Maintenance

Tractor storage
Before storing the tractor for an extended period of time, the following considerations should be taken into account.

Cleaning the tractor
Grease all the grease nipples.

Drain engine and transmission oil and fill with new oil.
Empty the fuel tank and pour into it one two gallons of fuel of the type used for calibrating pumps.
Run the engine for 10 minutes to ensure that the fuel is well distributed throughout the system.
Raise the hydraulic system arms and hold them up with supports.
Remove the battery and store it in a dry, warm place. Recharge it periodically.
Raise the axles, placing supports under them.
Empty radiator and engine block
Covering the exhaust

Preparing the tractor for use after prolonged storage.

Inflate tyres to the correct pressure
Refill the cooling system and the fuel tank.
Installing a fully charged battery
Check all oil levels
Remove the exhaust cover (not the rain cover).
Start the engine and check that all controls are working properly.
Drive the tractor without load to make sure it is working properly.

Alternator charging system

To avoid damage to the alternator charging system components, the following operating precautions must be observed:
Never make or cut connections to the charging circuit, including the battery, while the engine is running.
Never cross any load components to ground
Always disconnect the battery ground cable when installing or removing the alternator Always disconnect the battery ground cable when charging the battery on the tractor with a charger.
Never use an auxiliary battery of more than 12 volts.
Always, maintain correct, polarity when mounting the battery or use an auxiliary battery to start the engine.

CONNECT POSITIVE TO POSITIVE POST AND WIRE AND NEGATIVE TO NEGATIVE POST AND WIRE.

Preventive maintenance rules for a combine harvester or combined harvester

The combi-machine is an expensive machine, as are its repairs and spare parts. For this reason, it is recommended that the maintenance rules mentioned here be followed to the letter:

1. - Do daily

    a) Air filter (cleaning)

Engine oil
Hydraulic system
Transmission
Battery water (if required)
Radiator water
Motor fan belt voltage
Chain and belt tension
Tightening of wheel nuts
Tire air pressure
Retraction system

    b) Fill up the fuel tank after each working day at home
    c) Drain fuel filters
    d) Grease all your points
    e) Straightening and replacing guards and blades
    f) Cleaning the concave grids
    g) Clean the trap of stones

2. Do every 5 days

    a) Check oil level of final drives
    b) Check the oil level of the blade oscillating mechanism.
    c) Check oil level of the cylinder speed reducer mechanism.

3. Do every week
a) Lubricate the generator with two drops of oil.
b) Tighten all nuts and bolts.
c) General cleaning of the entire machine
4. Check and clean the engine breather every month.
5. Do every 600 hours
a) Check and calibrate injectors
b) Check and calibrate aiming devices
c) Check and flush the fuel tank

Periodic Maintenance Chart

Oil changes	Suitable oil	Change every
Engine	Oil number 30	100 hours
Transmission	Oil number 90	500 hours

Hydraulic system and filter	JD 303 Mobiloil, TEXACO TDH Fluid Type "A" automatic transmission fluid.	500 hours
End controls	Oil number EP-90	250 hours
Cylinder speed reduction mechanism housing.	Oil number 90	Complete only
Fuel filter changes.		100 hours
Engine oil changes		100 hours

Finally, it is suggested to consider the operator's manual that each tractor dealer provides to anyone who purchases an agricultural tractor.

XIII.   The following service control is suggested

Combined model...
Engine number..
Serial number...
Working hours..

Services Hours
Engine oil change...............................
Hydraulic system oil change ....
Transmission oil change ....
Oil change final drives................................
Fuel filter changes................................
Oil filter changes...............................
Oil changes address ....
Calibrate injectors................................
Calibrate the aiming device................................
Flush fuel system................................
Wash cooling system...............................
General lubrication................................
Next service to ....

REMARKS

Date
....
Personnel who performed the service

XIV. Operation of a combine harvester, threshing machine or grain harvester

Harvesting is done with a threshing machine, combine harvester or combined grain harvester; action of cutting, threshing and separating the grain from the basic crop, when it has approximately 13% humidity.
Although in this case, it is not used properly a tractor, the combined has a 6 cylinder engine, then has up to 473 H.P. and with 13.5 liters, which requires special adjustments throughout its mechanism of cutting, threshing and separation, are functions performed in a single action, to obtain clean grain and residues of the stalk and leaves, the head has a considerable width of advance of up to 45 feet, which is equivalent to 13.71 meters wide cutting or advance.

The suggested speed is 6 to 10 km/hour, and the forward width in meters is 13.71 and is obtained as follows, formula to convert feet to meters = feet/3.2808, therefore 45/3.2808 = 13.71 meters and that is a very adequate forward width and will finish the harvest in an adequate time.
The maintenance already mentioned in this document must be considered and the wear life is 2 000 hours.

This machine, are agricultural use is with own transmission and self-propelled by an engine that is usually six pistons and can generate up to 473 H.P., the operations that can perform in a certain time are: Cutting, harvesting, feeding, threshing, separating and cleaning.

Parts and operation

Internal combustion diesel engine with four strokes and six pistons: Its function is to generate the necessary power to move the whole mechanism that integrates the threshing machine, combine harvester or combined and that are detailed below.

Windlass: Its function is to gather the plants of the harvest and direct them to the mowing unit, so that it makes the cut.

Mower unit or blades: Performs the cutting of the crop. If it is cut below the machine is fed with a lot of straw, creating problems of overfeeding, the cut should be such that it collects all the grain, but not all the straw.

Worm worm of the platform: It has a continuous spiral shape and its function is to transport the harvest from the ends to the center of the platform, so that later the feeder assembly leads them to the cylinder and concave.

Elevator or feeder chains: Its function is to transport the harvest to the cylinder and concave, so that these, in turn, make the threshing.

Cylinder and concave: They thresh the grain or separate the grain from the plant, by means of the rubbing action of the bar of the cylinder in its rotation with the concave. This is stationary and open bottom, so that the grain passes to the staggered tray.

Beater: Its function is to slow down and direct the straw towards the straw remover that comes from the cylinder and concave.

Fan: Provides a continuous, controlled flow of air through the sieve box area to separate the trash from the grain in the final cleaning operation.

Staggered tray: It receives the grain that falls from the concave, product of the threshing and directs it to the sieves.

Grain elevator: Transports the grain to the storage tank.

Straw removal: conveys the straw out of the machine

Re-threshing cylinder: Its function is to make a second threshing when it tends to pass harvest with tips or tails that were not threshed in the cylinder and concave.

Collecting tray: Collects the grains that fall through the open bottom of the straw removers and directs them to the upper sieve.

Sieves: This is where the final grain cleaning operation is carried out. The agitator set, which is precisely the drawer with the sieves, and the controlled flow of air from the fan are used to separate the grain from the trash, which comes out of the back of the machine.

Grain tank: Its function is to store the clean grain.

XV. Off-season maintenance of a combine, threshing machine or grain harvester.

It is very common that after the end of the harvesting season the combined harvester does not receive maintenance, because it is immediately stored. If when the next harvesting cycle starts, problems occur when the machine is used, belts, bearings, chains break, engine and hydraulic system failures, corrosion occurs, the combine settings are lost, etc. This means that the combine needs off-season maintenance.
This type of problem wastes time and large amounts of money for the farmer and to avoid them we recommend the instructions detailed here on off-season maintenance.

Cleaning

It is important to clean the harvester carefully before any off-season maintenance. Accumulated dirt immediately absorbs moisture, accelerates corrosion and even harbours harmful insects.

It is recommended that neither pressure washing nor water be used to clean the inside of the combine harvester. It would be ideal if the cleaning is done using compressed air when, when this is not available, it is sufficient to clean it as best as possible by hand.

It is recommended to start cleaning operations as follows: first remove all dirt and chaff from the cutting table, cylinder, concave, and so on, sieves, sieves, fan, regrinding cylinder (if any), lifting chains and storage tank.

Protection of moving parts. Unprotected metal surfaces

RECOMMENDED ANTICORROSIVESMobiloyl
66 PH-40
Texaco Rust Proff Compound "L" Texaco Rust Proff Compound "L"
Texaco Rust Proff Compound "L

BLADES AND CUTTER BAR

Remove the blade and treat it, together with the cutter bar, with petroleum or any other solvent, then store it in an oil bath. You can use for this purpose, used oil from engines, or in its effect use an antioxidant. Excepted here are the chains elevating the grain to the tank, because the pallets or rubber slats deteriorate if they are immersed in oil.

Afterwards, these implements can be hung in a suitable place for preservation.

MAIN HOIST

The elevator chain that transports the crop to the cylinder and concave, should be sprayed with a rust inhibitor without removing it, at the same time, it is necessary to loosen the tension of the same. Special attention should be paid to the application of rust inhibitor to the shafts of the variable speed pulleys, to avoid corrosion of the polished surfaces. The pulley faces must be clean and dry.

CYLINDER AND CONCAVE

After cleaning, it is advisable to spray them with an antioxidant.

BANDS

It is advisable to remove all rubber bands; hang them in a cool place and make sure they are well cleaned, as grease, oil or fuel can damage them.

CHAINS

All chains should be removed, washed and brushed with oil, dried and stored in an oil bath. Burnt oil can be used for this purpose. This operation is applied to chains that are accessible and sprayed with a rust inhibitor.

THRESHING CLUTCH

The clutch to connect the entire threshing system must be kept disengaged, otherwise the belt will be damaged and may stick.

JOINTS

All points of rotation of the various control joints, brakes, hydraulic controls, hand levers and pedals should be treated with rust preventative. Likewise, all belt tensioners, chains, jousting screws etc. should be rustproofed.

HYDRAULIC SYSTEM

The hydraulic system of the harvesters requires some attention, such as: emptying the main tank, cleaning or renewing the tank with new oil of the type recommended by the manufacturer and repairing leaks if any, adjusting the belt and repairing leaks if any, adjusting the belt that drives the hydraulic pump and testing the system several times to check that it is in good working order. Hydraulic cylinders should be stored in the closed position; if not possible, they should be treated with a rust preventative.

CLUTCH

The clutch does not require so much attention, but it is advisable to check its adjustments, especially the pedal, so that the gears enter without difficulty and at the same time detect any defect that the clutch may have.

TRANSMISSION, FINAL DRIVES, ENGINE

For maintenance of the transmission, steering, final drives and engine, it is recommended that you refer to the periodic maintenance chapters already reviewed above.

TYRES

The tyre pressure should be checked and the correct tyre pressure set. Finally, the machine must be shod so that the tyres are not in contact with the ground to prevent damage.

XVI.   Tractor systems that generate its movement

1.   Clutch system: Its function is to allow the selection of speeds and is composed of the following parts.
1.   - Disc and cover, which are attached by screws to the flywheel and the hole must be centered to engage the shaft of the gearbox.
2.   - Another part is the collar and fork, which must be coordinated so that when the clutch pedal is operated, the indicated speed is selected.

2.   Gearbox: Structure provided with special gears, which depends on which position should be selected the slow or fast lever or select the first, second, third or reverse speed, the gear that is located in the lower part, refers to the movement is transferred to the front wheels and when this system is operated, we say that this is applying the 4 x 4 system or four-wheel drive, which allows the tractor to be working to do so more easily. And it is equipped with a special oil for gearbox, to reduce friction between gears.

3. Transmission system: In this picture you can also see the system properly called transmission, which is also provided with very strong steel gears, which transfer movement to the rear wheels, each one is provided with structures called satellites, which allow the independent movement of each wheel, by means of the brake, that can be right or left, that is, if you press the pedal independently, either the right or left, will be working the free wheel turning to the corresponding side and this is very important especially when turning at the headlands, which is nothing more than the space needed for the tractor to turn and continue preparing the ground.

Another view of the gearbox and the rear part of what is properly called the transmission system.

View of the box of a transmission system under repair, note that, according to the design, traction is transferred to the front wheels, by means of the arrow on the right margin, which indicates that it is 4 x 4.

4. Hydraulic system: By means of a lever its function is to facilitate the lifting and lowering of any implement, which is coupled to the universal hitch system of three points, also allows the name of hydraulic steering, ie the steering wheel of the operator to turn the front wheels to the right or left is more comfortable, easy or simple and is provided with a special oil, to reduce friction between gears and stabilize the temperature.

It is usually located at the bottom of the operator's seat, and the pump can produce a maximum pressure of 70 to 140 kg/cm2.
Hydraulic system component, which is located on the underside of the operator's seat, inside the transmission system.

XVII. General explanation of how power is transmitted to the PTO, drawbar and rear tire movement.

Once the engine is running if the speed is neutral, the power generated is to the steering wheel, well, to operate the clutch system by pressing the clutch pedal selects the first or second speed in order that the tractor moves forward, and at that time is connected by an arrow and work both the gearbox and the transmission that is connected by gears, This last one is the one that makes the rear wheels move, by means of gears and satellites, so that each tire can be braked independently, when it is required, then the power take off can be connected, if you want it to work, coupling the implements that will be mentioned later and in the same way the type bar, which already has the optimum power, to pull the implements that will also be mentioned later. Also, once the engine is turned on, the hydraulic system is automatically activated, which allows the steering to move more easily and can lift and lower implements that are coupled to the universal three-point hitch system, as well as implements that are pulled by the drawbar, but connected to the hydraulic system by means of hoses.

XVIII. 4-stroke internal combustion engine agricultural tractor working and its relation with the implements that can make it work.

It is suggested to use a tractor of 95 H.P. to the steering wheel of the motor and that the power that is taken advantage of to the take of force is 81.9 h.p., to which we can denominate to him that it is an efficient machine, due to the power that contains and that can make the majority of agricultural works of the fastest form and of quality, and this is due to the fact that the tractor in all its structure and conformation comes more reinforced.

Plow with three, four or more discs or moldboard plow for fallowing:

This implement is used to carry out the first work of land preparation called fallow, with this operation, the residues of the previous harvest are incorporated into the soil and the bottom layer of it is exposed to the action of the environment, thereby achieving the destruction of pests, fungi and weeds. By breaking the hard soil is obtained a looser surface that allows to carry out later practices, it has to be done 30 days in advance to the tracking and planting of the crop to be established and a soil depth of 25 to 30 cm.
The plow is coupled to the universal system of three points, two side arms and the compression bar that is located in the middle, so that the adjustment that is required, is in the sense that the plow is not tilted forward or backward, so that the fallow work does not allow the plow is not nailed forward or raised backwards because it would perform a correct job, it is proposed that the leveling is done in a space with concrete, so that the plow is well leveled.
A speed of 5 am 8 km/hour is proposed.
Maintenance: it consists of keeping the disc plow clean and greasing it before the working day, in the grease fittings and it has a life of wear of 2,500 hours, that is to say, the discs will have to be changed or the whole plow will have to be replaced if it is the case.

Lifting harrow with 18 or 20 discs or, if necessary, a harrow with a puller for tracking:

This implement can be of 18 or 20 discs of lifting or in its case a harrow of pull which has a greater width of advance, the first is coupled to the universal coupling of three points and in the case of the harrow of pull this is coupled to the drawbar and coupled the hydraulic system by means of hoses, to be able to raise the mechanism including the discs and it is to the floor the two tires to be able to move the tractor and implement towards another part of land. So the tracking or second work to be done to establish a cyclical crop (less than a year), such as corn, sorghum, rice, wheat etc.. It is suggested a harrow with a greater width of advance as can be the harrow of pull, with the purpose of carrying out in the best way the work of tracking which should be done after 30 days that have passed after the fallow and is a work prior to planting, so it is necessary to give a step of harrow to 15 centimeters deep in order to undo the clods that remain after the fallow. A well prepared soil that includes the harrowing favors the germination of the seed, at the same time that eliminates the resistance that may exist for the growth of the roots, makes available to the plant the necessary nutrients. We suggest a speed of 6 to 10 km/hour.

for its development and production, ensures air circulation and retains more water for the benefit of the agricultural crop.
For its leveling, it is the same procedure as for the plow and it is also suggested that it be in a concrete part, for a better effectiveness in the leveling. Maintenance: it consists of keeping the harrow disks clean and greasing it before the working day, in the grease fittings and it has a life of wear of 2,500 hours, that is to say, the disks will have to be changed or the whole harrow will have to be replaced if it is the case.

Sowing machine:

If the crop is basic and does not require space between plants, a normal or traditional seeder is used, but if you are going to establish a crop that requires a certain distance between plants and whose destination is for the acquisition of grain, such as corn, sorghum, wheat, rice or beans, a precision seeder is used, ie deposits one or two seeds every certain distance, the mechanical seeder performs three tasks in a single action: furrows, fertilizes and sows.

As the universal three-point system is attached, it is suggested that the leveling be similar to the previous implement, on a concrete part. A speed of 6 to 10 km/hour is suggested.
Maintenance: it consists of keeping the seeder clean and greasing it before the working day, in the grease fittings and it has a wear life of 1,200 hours, that is to say, the selection plates will have to be changed or the whole seeder will have to be replaced if it is the case.

Cultivator:

First cultivation, with a cultivator coupled to an agricultural tractor, work that should be done 20 to 30 days after planting, the function of this work is to eliminate weeds and also rotate the soil to conserve maximum moisture. As the cultivator is coupled to the universal coupling of three points, the leveling is suggested in a concrete part and the more is its width of advance, the greater will be its effectiveness and it is proposed a speed of 6 to 15 km / hour.

Maintenance: it consists of keeping the cultivator clean and greasing it before the working day, the parts that require it and it has a life of wear of 2, 500 hours, that is to say it will be necessary to change the rudders or to replace all the cultivated one if it is the case.

Second Crop:

with cultivator coupled to the agricultural tractor, work to destroy the weeds that emerged after the first crop and keep the maximum soil moisture, this work should be done at 50 to 60 days after planting.
The same way of levelling and the same speed as in the case of the first crop is proposed.
Maintenance: it consists of keeping the cultivator clean and greasing it before the working day, the parts that require it and it has a life of wear of 2, 500 hours, that is to say it will be necessary to change the rudders or to replace all the cultivated one if it is the case.

First pest and disease control:

With sprinkler coupled to the tractor: during the grain crop cycle, when required. For this case the leveling is simpler, that is to say that it is not forced to one side or another side or forward or backward, as well as to do it in a space with concrete.
Suggested speed is 5 to 7 km/hour.
Maintenance: consists of keeping the sprinkler pump clean and greasing it before the work day, the parts that require it and has a wear life of 2,500 hours, i.e. you will have to change the hose or part of the mechanism or replace the entire sprinkler pump if it is the case.

Second pest and disease control:

With sprinkler coupled to the tractor: during the grain crop cycle, when required. For this case the leveling is simpler, that is to say that it is not forced to one side or another side or forward or backward, as well as to do it in a space with concrete.
Suggested speed is 5 to 7 km/hour.
Maintenance: consists of keeping the sprinkler pump clean and greasing it before the working day, the parts that require it and has a wear life of 2 500 hours, i.e. you will have to change the hose or part of the mechanism or replace the entire sprinkler pump if it is the case.

XIX. Other implements that can work the tractor equipped with a four-stroke internal combustion engine.

Stripping machine:

The function is the elimination of weeds, and this implement can be obtained with chains or blades for the elimination of weeds and is coupled to the universal three-point hitch, and also to the power take-off, so that it transfers movement at certain RPM (revolutions per minute), and this according to the make and model of the tractor, the leveling must be in a concrete part.
Maintenance: it consists of keeping the desvaradora clean and grease it before the working day, the parts that require it and has a life of wear of 2, 500 hours, that is to say it will be necessary to change the blades or chains that carry out the cut or part of the mechanism or to replace the whole desvaradora if it is the case.

Mill:

It must be coupled to the universal three-point hitch and the tractor PTO, it is required to level so that it can perform its crushing of plants or branches properly, it is suggested to consider the RPM.
Maintenance: consists of keeping the mill clean and greasing it before the working day, the parts that require it and has a wear life of 2, 500 hours, i.e. it will have to change the gears or blades that perform the crushing or part of the mechanism or replace the entire mill if it is the case.

Towing:

In this case it is only hooked to the drawbar and dragged, however, some trailers can be connected to the hydraulic system for greater safety, of course when loading with any product, care must be taken with the speed.
Maintenance: consists of keeping the trailer clean and greasing it before the working day, the parts that require it and has a life of wear of 2, 500 hours, that is to say, the tires or structure will have to be changed or replace the whole trailer if it is the case.

Levelling blades:

These can be coupled to the universal system of three points, are leveled and can move certain amounts of soil, sand or some other product, can come adapted to rotate through the hydraulic system to a certain position of degrees, in order to perform certain work and can also come designed tractors for leveling blades can work on the front of the tractor.

Maintenance: consists of keeping the leveling blade clean and greasing it before the work day, the parts that require it and has a wear life of 2,500 hours, and if it is the case, the parts that are used for leveling or to open a ditch or replace the entire leveling blade if it is the case.

Pruning implements for fruit trees with rounded locks:

Tool, which couples to the universal three-point hitch and the power take-off and works at certain RPM and directed by the hydraulic system, to have mobility and to be able to prune fruit trees, it is very common in mango, lemon or grapefruit.

Maintenance: consists of keeping the pruning shear clean and greasing it before the work day, the parts that require it and has a wear life of 2, 500 hours, and if it is the case, the parts that are used for pruning fruit trees will have to be changed or replaced if it is the case.

Manure or dust distributor:

Normally they are coupled to the universal three-point hitch and the PTO, in such a way that it transfers movement to the mechanism and the product can be applied in a circle to the ground, it is suggested to consider the RPM at which the PTO must work.
Maintenance: consists of keeping the manure distributor clean and greasing it before the work day, the parts that require it and has a wear life of 2,500 hours, and if it is the case, the parts used for manure distribution will have to be changed or replaced if necessary.

Roll or square balers.

This implement allows baling the forage already in rows and dry, either in rolls or squares, for this it is necessary that the baler is coupled to the universal three-point hitch, leveled and the power take-off at the corresponding RPM.
Maintenance: consists of keeping the baler clean and greasing it before the work day, the parts that require it and has a wear life of 2, 500 hours, and if it is the case it will have to change the parts that are used to pack or change it if it is the case.

Windrowers:

Implement which, once the forage is cut, allows to organize the crop in rows and once it is turned over and with certain humidity it can be baled, the correct leveling and speed in km/hour is proposed.

Maintenance: consists of keeping the swather clean and greasing it before the working day, the parts that require it and has a life of wear of 2, 500 hours, and if it is the case will have to change the parts that are used to align the cut crop or change it if it is the case.

Normal blinders:

They cut the forage crop and leave the crop parts in disarray, then the windrower fulfills its function of organizing it into windrows, it is suggested to consider leveling and the correct RPM and speed.
Maintenance: it consists of keeping the normal blinding machine clean and greasing it before the working day, the parts that require it and it has a life of wear of 2 500 hours, and if it is the case, the parts that are used to cut the established crop will have to be changed or replace it if it is the case.

Conditioning blinder:

Cut the crop, but organize it in rows, so that once with the right humidity, it can already be packed, suggesting the correct leveling, speed and RPM.
Maintenance: consists of keeping the blinder conditioner clean and grease it before the workday, the parts that require it and has a wear life of 2,500 hours, and if it is the case will have to change the parts that are used to cut and spin the crop established or change it if it is the case.

Forage harvesters:

These implements are coupled to the universal three-point hitch and the power takeoff, so it must be level, the speed must be adequate, in addition to the RPM, is to crush for example green corn, in milky state - doughy, which is when most of the nutrients are concentrated in fresh and livestock can take advantage of the best way, usually the distance between rows is less, so you have harvesters of forage corn one, two, three or four rows.

Maintenance: consists of keeping the forage harvester clean and greasing it before the working day, the parts that require it and has a wear life of 2 000 hours, and if it is the case it will have to change the parts that are used to harvest the forage crop established or change it if it is the case.

XX. Bibliography

Acevedo López Arturo, 1976, Combines (Combine harvesters), Chapingo
Archie A. Stone & Harold E. Gulvin, 1982. Stone & Harold E. Gulvin, 1982, Agricultural Machinery, Editorial C.E.C.S.A. Decima segunda impresión, Mexico.

Cuiris Perez Heliodoro, 2006, Course on Mechanization of Agricultural Production, Faculty of Agrobiology "Pdte. Juarez" UMSNH.

Donell Hunt, 1983, Agricultural Machinery, Economic performance, costs, operations, power and equipment selection, Editorial LIMUSA S.A., First edition, Mexico.

Roberto Rivas Valencia's own experience as a mechanic from 1969 to 1975, diesel mechanic shop called Inyección Diésel del Cupatitzio, owned by Mr. Enrique García Molina, and with the master mechanic Rubén Rodríguez Ponce, located in Zacatecas number 87, Col. Ramón Farias de Uruapan, Michoacán.

García Fernández José and García del Caz Rafael, 1983, Maquinas Agrícolas, Editorial Boixareu, 2nd Edition.

https://www.metric-conversions.org/es/longitud/pies-a-metros.htm (20/03/2021) Norton L. Roberto, 2012, Machinery Design, fifth edition, Mc Graw Hill Publishing House.

Rivas Valencia Roberto, 2011, Notes of Agricultural Machinery, authorized by the Honorable Technical Council of the Faculty of Agrobiology "Pdte. Juarez" of the UMSNH.

Rivas Valencia Roberto, 2019, Mechanized Financial Planning in Agricultural Management. Article iSsN(Online): 2347-3002. UMSNH.

Robles, Sanchez Raul, 1981, Grain and Forage Production, Second Edition, Limusa Publishing House

SARH-INIA del Norte-Centro, Experimental Agricultural Field Zacatecas, 1982

SARH-INIA, Guía para cultivar la cebada en el Bajío, 1982).

SEP, 2003, Motores Agrícolas, Editorial Trillas Octava Reimpresión.

SEP, 2008, Elements of Agricultural Machinery, Reprint 2008.

Soto Molina Saúl, 2011, Introduction to the Study of Agricultural Machinery, Editorial Trillas, Reprint 2011.

Ford 5600, 6600 and 7600 Tractors, 1985

Kubota tractors, March 2021, from Uruapan, Michoacan, Mexico.

Printed by Books on Demand GmbH, Norderstedt / Germany